THE COMPLETE GUIDE TO AUTOMOTIVE WIRING AND ELECTRICAL SYSTEM

A Comprehensive Overview of Automotive
Wiring and Electrical Systems

Edwin S. Richard

Disclaimer

This book is designed to provide information about automotive wiring and electrical systems. Every effort has been made to make this book as complete and accurate as possible, but no warranty or fitness is implied. The information is provided on an "as is" basis. The author and the publisher shall have neither liability nor responsibility to any person or entity with respect to any loss or damages arising from the information contained in this book.

ABOUT THE AUTHOR

Edwin S. Richard is a keen motor enthusiast with a plethora of knowledge in automotive systems. Based in Detroit, Michigan, the automotive center of the world, Edwin has devoted his career to studying and developing the mechanics that power our automobiles.

With a degree in mechanical engineering, Edwin has worked for numerous renowned automotive firms in the Detroit region. His time with these organizations enabled him to obtain direct knowledge with numerous vehicle systems, such as engines, gearboxes, suspension, and brakes.

Driven by his enthusiasm for autos, Edwin has made it his duty to remain up to speed with the newest breakthroughs in automotive technology. Whether it's researching the complexities of hybrid engines or investigating the possibility of electric cars, Edwin is continuously seeking fresh information to expand his grasp of automotive systems.

In addition to his professional skills, Edwin likes sharing his insights with others via writing. He has published articles to several automotive periodicals and websites, offering important insights and guidance to other vehicle lovers. Edwin's writing style is instructive and fascinating, helping readers to readily comprehend difficult automotive topics.

As a seasoned automotive specialist, Edwin has established himself as a reliable source of knowledge in the industry. His enthusiasm for autos, along with his comprehensive understanding of automotive systems, makes him a vital addition to the field of automotive engineering and technology. Whether it's debugging a mechanical problem or investigating the newest innovations in vehicle design, Edwin's skill and enthusiasm make him a renowned and sought-after automotive specialist.

When he's not working on automobiles or writing about them, Edwin likes visiting automotive events and car shows, where he can indulge in his enthusiasm for historic and high-performance vehicles. He is an active member of many automotive groups and associations, where he likes networking with other vehicle lovers and sharing ideas.

Overall, Edwin S. Richard's enthusiasm for autos and his thorough grasp of automotive systems make him a well respected expert in the industry. With his skills and experience, Edwin continues to contribute to the growth of automotive engineering and inspire others with his enthusiasm for all things automotive.

TABLE OF CONTENTS

INTRODUCTION

The history of vehicle wiring and electrical systems has been lengthy and complicated. It started in the early days of the automobile, when the internal combustion engine was first developed. At that time, wires were used to connect the engine to the various components of the vehicle, such as the braking system, the ignition system, and the electrical system.

For many years, the wires were made of copper and were insulated with rubber or fabric. As the technology improved, more efficient and durable materials were developed. In the early days, the wiring was labor-intensive and required a great deal of skill and expertise.

Over time, automotive wiring and electrical systems have become much more complex. Today, they are powered by multiple electronic systems, such as the engine control unit, the transmission control unit, the airbag control unit, and the anti-lock braking system. In addition, modern cars are equipped with various sensors, such as those used for navigation and entertainment systems.

To keep up with the advancements in technology, automotive wiring and electrical systems must be regularly inspected and maintained. This includes checking for loose connections, frayed wires, and faulty components.

The automotive wiring and electrical systems of today are crucial for the safe and reliable operation of your vehicle. Without them, your car won't be able to run properly and could potentially be a hazard on the road. Keeping your wiring and electrical system in good condition is essential for your safety, and that of your passengers.

Do you want to know more about automotive wiring and electrical systems? Then this is the perfect book for you. Automotive Wiring and Electrical Systems is an in-depth, comprehensive guide to the electrical systems that power today's cars, trucks, and other vehicles. Written with the experienced and novice mechanic in mind, this book provides an easy-to-follow, step-by-step approach to understanding the complex electrical systems found in today's vehicles.

This book covers topics such as the fundamentals of electrical wiring and components, troubleshooting techniques, and more. It also introduces you to the latest technologies in automotive electrical systems.

The book is packed with helpful tips, tricks, and illustrations, making it easy to understand the basics of automotive wiring and electrical systems. It also provides clear, concise instructions on how to diagnose, repair, and maintain your vehicle's electrical system. With this book, you'll be able to confidently tackle any electrical issue your vehicle may have.

Whether you're a novice or experienced mechanic, Automotive Wiring and Electrical Systems is the perfect guide for understanding the electrical systems that power today's vehicles. With this book, you'll be able to confidently diagnose, repair, and maintain your vehicle's electrical system.

CHAPTER 1: WHAT IS AN AUTOMOTIVE ELECTRICAL SYSTEM?

In vehicles, particularly the current ones, types of components are electronics and function electrically. Well, charging systems are the fundamental electrical system of a car which comprise an alternator, battery, and voltage regulator. These components provide a source of electricity to other electrical components in the car.

Although voltage regulators are incorporated in the alternator which functions as energy converters. There are hundreds of electrical components that depend on the electrical system of the car.

DEFINITION OF VEHICLE ELECTRICAL SYSTEM

Vehicle electrical systems are electrically operated equipment in a vehicle, they collect energy from the battery and return it to the battery via the hearth. The charging system includes an alternator and battery. This battery is used to power the starting motor and assists the engine to start running while the alternator is used to charge the battery and other electrical components in the vehicle.

Apart from this charge, certain automotive vehicles are constructed with magneto ignition which creates power that drives a spark plug in the combustion chambers. It's also utilized to power various electrical components, which helps to preserve battery power. Although some ignition mechanisms rely on the battery's power. All electrical circuits in automobiles are opened and closed either by switches or relays and fuses are employed to keep them from overloads.

Applications

The major function of the electrical system is to power all electrical and electronic components in a vehicle. beginning with the electrical motor, sensors, gauges, heating element, headlights, braking, and trafficator lights, radio, television, air conditioning system, blowers, interior lights, refrigerator system, ignition system, etc. All these components get electricity from a battery and the battery is charged by the alternator.

Note, while the engine is operating all electrical gadgets are powered with the alternator regulator's power. This is because the alternator output is larger than the battery current while the engine is operating.

Functions

The following are the functions of the vehicle's electrical system:

1. The fundamental role of a vehicle's electrical system is to create, store, and provide electrical current to the different electrical equipment in an automobile.

2. It runs all electrical parts/components of a vehicle.

3. Again, the vehicle's electrical systems aid in maintaining gadgets in excellent functioning condition as they can accomplish various characteristics.

Components Of Car Electrical System

Magneto: A magneto ignition system or high-tension magneto is an ignition system that employs magneto to provide high voltage for the creation of energy. The power produced is further utilized to operate cars and other electrical components in the system. The magneto is a mixture of a distributor and generator created as one unit, making it distinct from the ordinary distributor that produces spark energy without external voltage.

There is a sequence of revolving magnets that break an electrical field, creating an electrical current in the coil's main windings. The current charge will then double as it moves to the coil's secondary windings. This is because there are many more times the number of windings in the secondary circuit compared with the one in the main circuit, which then causes the multiplied charge magneto to generate a spark at a greater voltage than was formed in the primary windings.

Alternator: An alternator is one of the essential and unavoidable pieces of a car charging system as it performs the finest function. The electrical power that charges the battery comes from the alternator, although the current generated is alternating current (AC). This AC electricity is quickly converted to direct current (DC) since autos utilize a 12-volt DC electrical system. A dead battery does not indicate there is anything wrong with it. It is only that being deprived of charge, this is why the alternator is also examined if a vehicle is not starting.

Voltage Regulator: The voltage regulator regulates the alternator's power output. This device is commonly situated in the alternator since it controls the charging voltage that the alternator generates. It maintains a voltage between 13.5 and 14.5 volts to safeguard the electrical components in the car. in current automobiles that employ ECU to recognize when the battery needs to be charged as it regulates the voltage provided. The warning bulb on the dashboard signals something is amiss with the charging system. Often the warning bulb signals a malfunctioning alternator, resulting in an uncharged battery.

Battery: A battery is another crucial component of a car charging system since it acts as a reservoir of electrical power. An engine starting motor is directly linked to the positive terminal. It assists with turning the part making the engine start

As the engine is operating, the alternator instantly charges the battery. The battery may also deliver electricity to the electrical components while the engine isn't operating.

How Does An Automobile Electrical System Work?

The electrical system of an automobile is a closed circuit with an independent power source, the battery. It runs on a minuscule fraction of the power of a residential circuit.

A TYPICAL ELECTRICAL SYSTEM
Apart from the primary charging, starting, and ignition circuits, there are various circuits that power lights, electric motors, the sensors and gauges of electronic instruments, heating elements, magnetically

driven locks, the radio, and so on. All Circuits are opened and closed either by switches or by relays - remote switches driven by electromagnets.

Current runs down a single wire from the battery to the component being powered, then back to the battery through the car's metal body. The body is linked to the ground connection of the battery via a stout wire.

Earth-return System

In a negative (-) earth-return system, the current flows from the positive (+) terminal to the component being operated. The component is earthed to the automobile body, which is earthed to the negative (-) terminal of the battery.

This sort of circuit is termed an earth-return system; any section of it linked to the automobile body is said to be earthed.

The intensity of the current is measured in amperes (amps); the pressure that pushes it around the circuit is termed voltage (volts). Modern automobiles feature a 12-volt battery. Its capacity is measured in amp/hours. A 56 amp/hour battery should be able to supply a current of 1 amp for 56 hours, or 2 amps for 28 hours.

If the battery voltage declines, less current flows, and finally there is not enough to make the components operate.

Current, Voltage, And Resistance

The amount to which a wire opposes the passage of electricity is termed resistance and is measured in ohms.

Thin wires conduct less readily than thick ones since there is less area for the electrons to flow through.

The energy required to drive current through a resistance is turned into heat. This may be advantageous, for example in the extremely thin filament of a light bulb, which shines white hot.

However, a component with a high current consumption must not be connected using wires that are too thin, otherwise, the wires may overheat, blow a fuse, or burn out.

Every one of the electrical units of estimation is interrelated: a tension of 1 volt makes a flow of 1 amp course through an opposition of 1 ohm. Volts divided by ohms equal amps. For example, a light bulb with a resistance of 3 ohms, in a 12-volt system, uses 4 amps.

This implies it must be linked using cables thick enough to handle 4 amps easily.
Often the power consumption of a component will be expressed in watts, which are calculated by multiplying amps and volts. The light in the case uses 48 watts.

Positive And Negative Polarity

Electricity flows from a battery in one way only, and certain components operate only if the flow through them is in the proper direction.

This acceptance of a one-way flow is termed polarity. On most automobiles, the negative (-) battery connector is earthed while the positive (+) one supplies the electrical system.

This is termed a negative earth system, and before purchasing an electrical accessory a radio, for example, ensure that it is of a kind acceptable for your car's system. Fitting a radio with the wrong polarity may destroy the set, however, most vehicle radios include an external switch for adjusting the polarity to fit that of the automobile. Switch to the right setting before fitting.

Short Circuits And Fuses

If the wrong-sized wire is used, or if a wire gets cut or disconnected, this might generate an unintended short circuit that bypasses the resistance of the component. The current in the wire may get dangerously high and melt the wire or create a fire.

The fuse box is commonly positioned in a cluster of components, as depicted above. The box is seen with the cover removed.

To prevent this, supplementary circuits contain fuses.

The most common sort of fuse is a small length of thin wire wrapped in a heatproof housing, generally glass.

The size of the fuse wire is the thinnest that can carry the typical current of the circuit without overheating, and it is rated in amps. The quick rush of high current in a short circuit causes the fuse wire to melt, or 'blow', breaking the circuit.

When this occurs, examine whether there is a short circuit or a disconnection, then install a new fuse with the right amperage rating (examine Checking and replacing fuses).

There are several fuses, each safeguarding a small set of components so that one blown fuse does not shut down the entire system. Many of the fuses are clustered together in a fuse box, however, there may also be line fuses in the wiring.

Series And Parallel Circuits

A circuit frequently involves more than one component, such as bulbs in the lighting circuits. It depends on whether they are linked in series one after the other or in parallel side by side.

A headlight bulb, for example, is intended to have a degree of resistance so that it requires a specific current to shine properly. But there are at least two headlights throughout the circuit. If they were linked in series, an electric current would have to run through one headlamp to get to the other.

The current would contact the resistance twice, and the twofold resistance would half the current, such that the lights would shine only feebly.

Connecting the lights in parallel means that energy travels through each bulb only once.

Some components must be linked in series. For example, the sender in the gasoline tank modulates its resistance according to the quantity of fuel in the tank, and 'sends' a little electrical current to the fuel gauge.

The two components are linked in series such that the fluctuating resistance in the sender will impact the position of the needle on the gauge.

Ancillary Circuits

The starting motor has its hefty wire, straight from the battery. The ignition circuit gives the high-tension impulses to the sparkplugs; and the charging system contains the generator, which recharges the battery. All the other circuits are termed auxiliary (subsidiary) circuits.

Most are connected via the ignition switch so that they operate only when the ignition is turned on.
This stops you from mistakenly leaving anything powered on which can cause the battery to run flat.
The side and tail lights, however, which you may need to leave on while the vehicle is parked, are always wired independently of the ignition switch. while adding additional accessories, such as a rear window warmer which uses a significant current, always connect it via the ignition switch.

Some ancillary components may be activated without the ignition turned on by flipping the switch to the 'auxiliary' position. A radio is generally hooked via this switch, so that it may be played with the engine off.

Wires And Printed Circuits

The instrument connections to this printed circuit are disconnected by pressing the inbuilt hooks on either end.

Wire and cable diameters are characterized by the maximum amperage that they can carry safely.

A complicated network of cables runs through the automobile. To minimize misunderstanding, each wire is color-tagged (but just inside the car: there is no national or worldwide system of color-coding).

The color coding, however, is a handy reference to tracing wiring.

Where wires run side-by-side they are tied together in a bundle, in a plastic or fabric sheath, to keep them clean and less difficult to fit.

This bundle of wires spans along the length of the automobile, with single wires or small groups of wires emerging as appropriate, and is termed the wiring loom.

Modern automobiles typically require room for numerous cables in small areas. Some manufacturers now employ printed circuits instead of bundles of wires, notably near the back of the instrument panel.

Printed circuits are plastic sheets on which copper traces have been 'printed'. Components are inserted directly into the rails.

CHAPTER 2: AUTOMOTIVE WIRING

The Basics

Unless you've been fixing or wiring 12-volt equipment for a long time, the world of automobile wiring might be frightening. You'll need to comprehend some jargon and technical concepts to get started. To ensure you're purchasing the correct wire for the task, here's a critical summary of what you need to know.

Wire Size/Gauge

Wire size is measured by gauge, and it's the reverse of what newbies may think: The thicker the wire, the lower its matching gauge number.

For instance, an 8-check wire is thicker than a 20-measure wire. The gauge you require will depend on the current demand of the circuit and the wire length between the circuit and the power source.

If you're shopping for repairs, make a point of having the technical manuals for any tractors, equipment, and other vehicles you maintain ready. The handbook frequently sets out precisely what gauge of wires are required for specific components, so you'll be able to have the appropriate wires in stock.

Also, bear in mind that wire of 8 gauge or thicker is often referred to as "cable."

Amperage

Next up is the "load" a wire can handle, which is stated in amperes, typically abbreviated to "amps." Amperage or amps give you the entire quantity of electricity that can travel through the wire. You should know the current rating of every wire in your shop and verify that it's correctly labeled. Overloading a wire is risky for both your team and your clients; a wire with an improper current rating might short out, stalling machinery in unsafe areas, destroying vital systems, and even igniting fires in severe circumstances. Know your amps, and always double-check that the automotive wire you're getting can carry the load.

Check to see what relays, fuses, circuit breakers, and other overload protection your wiring job needs, and similarly make sure you know what connections will be utilized for the task. Too many amps may short out connections, too.

The Fundamental Formula For Amps, Gauge, And Wire Length

In general, the longer your wire has to be run for the application—and the more power that needs to flow through it—the lower the gauge (i.e., the larger the thickness) needed.

Materials: Copper and Aluminum

Automotive wire is offered in two basic materials: copper and aluminum. Copper is more conductive, flexible, and less likely to corrode than aluminum; nevertheless, copper weighs more and is prone to price changes, so you'll need a reliable distributor to keep your prices reasonable. Aluminum, by comparison, is lighter and cheaper, but less robust and more prone to corrode or develop electrical resistance over time. Most automotive applications utilize copper wire.

Insulation

Automotive wire is commonly covered with either general-purpose thermoplastic (also called GPT or PVC) or cross-link insulation, so termed because the insulation material is extruded under heat and pressure, altering the molecular structure and enabling it to tolerate greater temperatures. GPT insulation is more cost-efficient, but cross-link insulation is more robust and heat resistant.

As a rule, look at how the wire is being utilized. Applications that are expected to confront exceptionally high temperatures like engine compartments should utilize cross-link wire; typical in-the-cabin applications should use GPT/PVC-insulated wire.

Cross-link insulated wire comes in three basic varieties, defined by insulation wall thickness.

Twisted Pair Cable for CAN Bus Applications

If you are dealing with automobile CAN bus wiring, make sure to acquaint yourself with twisted pair cables in their shielded and unshielded versions, which are crucial for avoiding electromagnetic interference with signals conveyed over the CAN bus system.

Wire Coloring, Striping, and Printing

Automotive wire is available in hues spanning the rainbow. One of the greatest methods to make regular maintenance and construction work simpler is to allocate a distinct color for each kind of automobile wire according to its final usage.

Colors are used commonly to identify and trace a wire to a related circuit. Tractor trailers, for example, utilize a conventional color scheme - brown wire representing the left light, yellow signifying the right light, and so on. Many market areas have their standard wire coloring methods. For example, consider wire color-coding standards in the maritime sector.

A further approach to build on your color scheme is to add a color stripe. For even more clarity you may have printed language that specifies precisely what the wire is for.

Most wire providers will be able to execute client striping and printing. Being able to sort wire fast helps you to be more efficient in building and maintenance, and lets you rapidly access and refill the wire you're using the most. It'll also enable you to color-code routine chores requiring wiring, freeing up more experienced personnel to deal with more intricate duties.

Connections

Wiring normally involves two kinds of connections: soldered and solderless.

Soldering is time-consuming and prone to mistakes and, with the large assortment of solderless connections available, is considered outdated for most automobile wiring today. Many wire and connection manufacturers oppose soldering owing to the complexity of providing quality control throughout the procedure.

Solderless connections have become the quickest and simplest method to terminate cables. They entail the use of readily accessible automotive wire connections such as spade terminals, ring terminals, and fast disconnects. Learning how to construct a correct crimp is vital for many of these items. Solderless connections may not be sturdy enough for many applications. However, you may equip yourself with extra strain relief materials like heat shrink or vinyl tape to ensure the lifespan of solderless connections.

The Characteristics And Types Of Automotive Wires

There are two types of car wires: high-voltage wire and low-voltage wire. Both are treated with copper multi-core cables. The cross-sectional area of the wire is largely determined according to its operating current.

For electrical appliances with tiny currents, to guarantee that the conductors have a specified mechanical strength, the cross-sectional area of the conductors should not be less than 0.5 mm2. Since the starter is short-term work, to guarantee that the starting can transfer adequate power when it operates regularly, the voltage drop created by the current per 100A on the line cannot exceed 0.1v-0.15v.

Therefore, the cross-sectional area of the wire employed is enormous. The high-voltage wire of the auto has an extremely high endure voltage, by and large, ought to be above 15kv, so its cross-sectional region is little (because the current is tiny), about 1.5mm2, the thickness of the insulation layer is much lower than that of the lower pressure line, and rubber insulation is used, plus Dipped cotton fabric.

Commonly utilized automobile models are national standard QVR-105, Japanese standard AV, AVS, AVSS, AVX/AEX, German standard FLRY-B, FLRY-A, FLRY-A, FLRY-B, FLRYW-A, FLRYW-B, beauty GTE, GPT, GXL, SXL, TWE, TWP, TXL.

The basis of selection of the automobile wire is defined by the degree of insulation of the wire, the current transmitted, and the needed mechanical strength. For long-term electrical equipment, 60% of the real current carrying capacity may be employed; for short-term working equipment, the actual carrying capacity can be between 60% and 100%.

The vehicle's electrical system is a low-voltage power source with considerable operational current and large voltage loss. Excessive voltage loss will impair the regular functioning of electrical devices. Therefore, when the conductor cross-section and selection, the voltage loss must not exceed a particular value: 12V system is not more than 0.5V, 24V system is not more than 1.0V. The actual operating current of the conductor must not be more than the permissible current-carrying capability of the conductor.

Wiring Harnesses

The wiring harness is the combination of electrical cables, or assemblage of wires, that links all electrical and electronic (E/E) components in the motor vehicle, such as sensors, electronic control units, batteries, and actuators. The wire harness controls the energy and information transfer inside the E/E system to perform main automotive duties, such as steering and braking as well as ancillary car functions, such as ventilation and infotainment.

Automobiles and other road vehicles such as lorries and buses are one of the most demanding applications for mechanical and electrical design. The electrical systems in these applications must perform correctly throughout a broad variety of environmental conditions, encountering severe temperature, humidity, sunshine, dirt, vibration, and more. The electrical system must also fulfill user expectations of dependability and necessary safety-critical design guidelines.

Automobiles include several wires and cables which would run over many kilometers if completely expanded. By tying them into a cable harness, they may be more protected against the undesirable effects of vibrations, abrasions, and wetness. Constricting the wires into a non-flexing bundle maximizes the utilization of space while lowering the chance of a short. Since the installer has just one harness to install (as opposed to several cables), installation time lowers, and the method may be readily standardized. Binding the wires into a flame-retardant sleeve also minimizes the danger of electrical fires. Designing such a contemporary vehicle electrical system is a crucial responsibility in the automobile development process and this is done in wire harnesses.

How is a wire harness created?

The electronic components within a car are rising day by day and offering novel issues in terms of managing the wire harnesses that link them.

A wire harness is a carefully constructed system that keeps multiple wires or connections organized. It is a systematic and integrated arrangement of cables inside an insulating substance.

The objective of the wire assembly is to transfer a signal or electrical power. Cables are linked together via straps, cable ties, cable lacing, sleeves, electrical tape, conduit, or a combination thereof.
Rather than manually routing and joining separate strands, the wires are trimmed to length, bundled, and then fastened to the

terminal or connector housing to create a single component.

The wiring harness is developed in two phases. It is developed in a software tool first and then the 2D and 3D layout is shared with manufacturing factories to produce the harness.
The exact method of automobile wire harness design contains the following steps:

First, the electrical system engineer offers the operations of the complete electrical system, including the electrical load and associated special needs. The quality of the electrical equipment, the installation location, and the manner of connection between the wire harness and the electrical equipment are all essential aspects.

From the electrical functions and needs supplied by the electrical system engineer, the full vehicle electrical schematic is constructed by adding components necessary for a function and linking them together. The functions which are widely utilized across various vehicles in an architectural platform are kept together.

After the schematic is determined, the wire harness design is produced. In one platform, the end customers might have a range of needs. It is incredibly time-consuming and costly if distinct designs are generated for each end user's needs independently. So, the designer takes care of the many versions when creating the wire harness.

At the conclusion, a 2D depiction of all the wiring designs is made to demonstrate the way various wires are bundled and how the bundles are coated to secure the wires. End connections are also depicted in this 2D illustration.

These designs may connect with 3D tools for the import and export of details. The wire lengths may be loaded from the 3D tool and the end-to-end connection information is exported from the wiring harness tool to a 3D tool. The 3D tool utilizes this data to add passive components such as straps, cable ties, cable lacing, sleeves, electrical tape, and conduits at suitable areas and send them back to the wire harness tool.

After the design is finalized in the software, the wire harness is made in the manufacturing facility beginning with the cutting area then the pre-assembly area, and lastly in the assembly area.

CHAPTER 3: IGNITION SYSTEMS

An ignition system is a mechanism used in various forms of an internal combustion engine, generally a petrol engine to ignite the fuel-air combination. This ignition is purposefully produced so that the explosion in the combustion chamber may be completed. This is to say, that the spark that happens in the ignition system (spark plug) causes the fuel-air combination to burn.

Just as previously indicated, the ignition system is included in spark-ignition internal combustion engines, however, it's also applied in some other mechanical applications. But it's fairly common on the gasoline engine. However, the procedure is different in compression ignition diesel engines as the fuel-air combination is ignited by the compression heat which causes the deletion of the spark plug. This is another subject of conversation you can check out below.

Induction coil systems

Magneto systems were overtaken by systems employing an induction coil when batteries began to appear more commonly in vehicles (as a consequence of the growing usage of electric starting motors). The 1908 Ford Model T and the 1886 Benz Patent-Motorwagen both had a tremble coil ignition mechanism, in which the shudder disturbed the current through the coil and created a flurry of sparks during each firing. At the correct moment throughout the engine cycle, the tremor coil would switch on. The four-cylinder engine of the Model T contains a trembling coil for each cylinder.

Distributor-based systems

Charles Kettering devised an improved ignition system at Delco in the United States, and it was initially utilized in Cadillac's 1912 automobiles. A single ignition coil, breaker points, a capacitor (to stop the points from pulling at break), and a distributor (to transfer the proper amount of current from the ignition coil to the right cylinder) comprised the Kettering ignition system. Due to its lower price and relative simplicity, the Kettering system became the standard ignition system in the automotive industry for many years.

The Function Of The Ignition System

Below is the purpose of an ignition system in spark-ignition internal combustion engines:

The basic purpose of an ignition system is to produce an electric spark in the engine combustion chamber at the right moment so that the fuel and air combination may ignite. It generates 30,000 volts across the spark plug.

A high spark voltage spark is provided to each spark plug in the precise order.
There is varying spark timing with load, speed, and other factors.

The spark is timed so it may occur when the piston is reaching the top dead center.

Applications Of The Ignition System

Below are the uses of the different kinds of ignition systems in vehicle engines:

The technology is utilized in two-wheeler vehicles (SI engines).

Just as the battery is used to create power in the battery ignition system, magneto is utilized for the creation of electricity.

Finally, the ignition system is extensively utilized in applications such as tractors, outboard motors, washing machines, marine engines, power units, and natural gas engines.

Types Of Ignition System

Below are the three major kinds of ignition systems used in spark-ignition internal combustion engines:

Magneto ignition system: In the magneto types of the ignition system. magneto acts as the key component needed for producing the energy of high voltage. This high voltage is then utilized for the creation of energy which is further used to operate automobiles. The system is a combination of a distributor and a generator designed as one unit. This makes it distinct from the usual distributor that produces spark energy without external voltage.

Electronic ignition system: The electronic varieties of ignition systems are controlled electronically and are powered by a battery, unlike the previous one that employed magneto. It contains negative and positive terminals; the negative terminal is grounded while the positive is attached to the ignition switch. So, when the switch is on, electricity is given to the electronic ignition module via the cables. The electricity is then transmitted to the ignition coil that has two winding; main winding and secondary winding. These windings are insulated and the main is thicker than the secondary winding. There is a rod between the windings that produces a magnetic field.

Battery ignition system: The battery versions of an ignition system are extensively employed in vehicles to create sparks using spark plugs with the help of a battery. It's frequently seen in 4-wheeler vehicles but is increasingly employed in two-wheeler vehicles that draw electricity from a 6-volt or 12-volt battery in the ignition coil.

Components Of The Ignition System

Below are the components of the various kinds of ignition systems and their functions:

Magneto ignition system

The components of the magneto ignition system are a magneto, distributor, capacitor, cam, contact breaker, and ignition switch. Their role has been detailed in the whole text.

Battery ignition system

The components of the battery ignition system batteries are the ignition switch, the ignition coil, and the ballast resistor. Its components also comprise a contact breaker, distributor, capacitor, and spark plug. Check the whole article to see their function.

Electronic ignition system

Components of the electronic ignition system also include a battery, distributor, capacitor, ignition control module, armature, ignition coil, and spark plug.

Working Principle

The operation of an ignition system is less sophisticated and may be simply comprehended. It is evident from the description of the above sections you are now acquainted with the functional elements and the functioning of the system.

Most of the kinds of ignition systems operate with batteries but few can produce power independently.

Advantages And Drawbacks Of Ignition System

Advantages:

Below are the benefits of the ignition system:

1. Less maintenance is necessary with magneto ignition systems, it's less costly, consumes less space, and does not require a battery. It offers great operating efficiency owing to a high-intensity spark and is less prone to mistakes since the battery is not utilized.

2. Another benefit of ignition systems is that the choice of electronic kinds has

fewer components and also requires little maintenance. Its efficiency is also excellent and it creates less pollution. Another benefit of the electronic ignition system is that it boosts fuel economy.

3. The benefit of battery kinds of ignition systems is that there is excellent intensity of spark. It also produces a high concentration of spark even when the engine speed is low or at the initial beginning. It also has minimal maintenance exactly like other kinds of ignition systems.

Disadvantages:

Despite the significant benefits of the ignition mechanism. certain limits still emerge. the following are the drawbacks of the ignition system:

1. The downside of the magneto type of ignition system is that there is a poor quality of spark at the initial start at low speed. Misfiring could also occur owing to leakage and the cost of the system is expensive.

2. Disadvantages of the electronic forms of ignition systems are that the cost of the system is significantly high and might take considerable space since a battery must be utilized to power the device.

3. In the battery, the downsides include periodic maintenance for the battery alone, greater space consumed, and efficiency reduced with a fall in spark intensity.

CHAPTER 4: CHARGING SYSTEMS

The vehicle is fitted with several electrical equipment to drive securely and pleasantly. The car demands power not just when going but also when it stops.

Therefore, the car contains a battery for a power source and a charging mechanism to create energy by the engine operating. The charging system gives power to all the electrical gadgets and charges the battery.

The Charging system is an integral aspect of the electrical system. It supplies electrical current for the lights, the radio, the heating, the engine's electrical systems, and other electrical accessories. It also keeps the batteries in a charged condition, recharging them as required.

The charging system includes three primary components: the alternator, the voltage regulator, and the batteries.

The alternator creates electrical power to drive accessories and refresh the batteries. It is generally powered by a belt situated off the crankshaft. Mechanical energy from the crankshaft is transformed by the alternator into electrical energy for the batteries and accessories.

The voltage regulator functions as an electrical traffic officer to manage alternator output. It knows when the batteries require recharging, or when the car's electrical demands grow, and changes the alternator's power appropriately.

The batteries are a supply of synthetic electrical power.

Their major role is to crank the engine. They also offer power to car accessories when the electrical demand is too severe for the alternator alone.

Three-phase alternating current (1) When a magnet spins inside a coil, a voltage will be formed between both ends of the coil. This will give birth to an alternating current.

(2) The link between the current produced in the coil and the location of the magnet is illustrated in the figure. The maximum amount of current is created when the N and S poles of the magnet are closest to the coil. However, the current flows in opposite directions with each half-rotation of the magnet. The current that creates a sine wave in this way is termed "single-phase alternating current".

Components And Functions

In general, the components of the charging system are composed of alternators and regulators. However, the charging system has to incorporate certain components so that the power produced can be provided to the battery and all electrical loads securely and correctly. The component consists of;

1. Battery

The role of the battery is as a store of electrical energy. Like a storehouse, the battery will store all the electrical energy created by the alternator and then this stored power is extracted as required.

2. Fuse and Fusible linkages

Fuse and fusible linkages have distinct purposes even though have the same form. The fusible link may be known as the primary fuse which is located near the battery positive terminal. The objective of this fuse is to safeguard the whole electrical system of the automobile against high currents. While the job of the fuse is as a safety of a series of particular electrical wires, in a normal charging system there are two fuse with identical capacity (around 10-15 Ampere). A fuse is utilized as a voltage regulator fuse and another fuse is used to safeguard the CHG and Voltage relay.

3. CHG Lights

CHG lamp often also called a "charging warning light" is an indicator light to show the existence failure of the charging system. When the ignition key ON then this light will light up properly, as well as when the engine life of this lamp should come on, if it is dead then it might signify the charging system failure.

4. Ignition key

The ignition key operates as a switch. The charging system will be triggered automatically while the engine is running, but producing a magnetic field on the rotor coil must be done via a switch.

An ignition switch is used as a switch to connect and disconnect electricity (positive battery currents) from the battery to the rotor coil. When the ignition key is ON, then the power from the battery to the coil rotor will be linked. However, when the ignition key is switched OFF the power supply will be shut off. So it is not feasible for the alternator to create electricity while the ignition key is OFF even when the engine crankshaft is revolved.

5. Regulator

The job of the regulator is to control the voltage produced by the alternator. Why should it be there? because the voltage created by the alternator affects the engine's RPM. This implies that if the engine RPM is low, the alternator voltage is also low, but if the engine RPM is high then the alternator voltage is likewise high.

The regulator will be used to maintain the voltage produced by the alternator from exceeding 14 volts even if the engine operates at high RPM. This voltage setting seeks to safeguard the electrical components of the car to avoid over-voltage.

There are two sorts of regulators, namely type or conventional type and type of IC.

The point type/conventional employs two coils to regulate the alternator's output voltage. While the IC Regulator employs an IC circuit (Integrated Circuit) to control the output voltage.

6. Altenator

The role of the alternator is to convert a partial engine's spinning energy into electricity. The alternator input originates from the engine pulley linked with a V belt, the rotation of the rotor will induce the crossing of the magnetic force line with the stator coil so that the electrons stream on the stator coil.

The power in the stator coil is not immediately linked to the battery, but it must travel via the diode bridge to rectify the current.

7. Charging Wires

The job of charging wires is to link every component of the charging system, there are at least two kinds of wires: standard wire and B + wire. The standard wire has a tiny diameter like the car's electrical wiring in general, the role of this wire is to connect each terminal to the full charging system.

While the B + wire has a greater diameter than the standard wire and virtually equals the starter wire. The role of this wire is to link the terminal B alternator with the Battery.

Working Principle

At the point when the start switch is in the ON position, current flows from the battery to the alternator.

The explanation behind this is as follows. The alternator commonly utilized for the car creates energy by spinning the magnet. The magnet is not the permanent magnet but the electromagnet that creates magnetic force by flowing energy within. Therefore, it is required to deliver energy to the alternator before starting the engine to prepare for producing electricity.

Ignition switch ON (while the engine is running)

Functions Of Alternator

The alternator plays a key part in the charging system. The alternator has three roles producing electricity, rectifying current, and controlling voltage.

(1) Generation

Transmitting the engine rotation to the pulley via the v-ribbed belt moves the electromagnetized rotor, creating alternating current in the stator coil.

(2) Rectification

Since the power produced in the stator coil is alternating current, this cannot be utilized for the DC electric equipment fitted on the car. To utilize the alternating current, the rectifier is needed to rectify the alternating current into direct current.

(3) Regulation of voltage

IC regulator adjusts the produced voltage to keep the voltage constant even when the alternator speed or the quantity of current flowing into the electric devices varies.

Charging System Diagnosis

The following basic information has been prepared as a help for charging system troubleshooting. Refer to the relevant Original Equipment Manufacturer's service manual for particular information related to charging system diagnostic procedures and safety considerations for your vehicle.

BENCH TESTING

If an alternator test bench is available, use the methods listed in the bench tester's instruction manual to conduct an alternator performance test. This test will establish whether the alternator output is within its performance standard, eliminating needless alternator replacement.

If the alternator output is within specification during bench testing, repair faults in the rest of the vehicle's charging circuit and other electrical circuits that may impact charging circuit performance. Refer to the relevant vehicle manufacturer's service manual for the techniques and circuit diagrams required to diagnose and rectify other charging circuit faults.

If the test bench findings indicate the alternator's output performance to be out of specification, replace the alternator. Follow the car manufacturer's suggested procedures to examine the rest of the charging circuit and any electrical circuits that may impact charging circuit performance.

CHAPTER 5: LIGHTING SYSTEMS

Various car lighting systems have a specific purpose: to aid the driver or inform other motorists. However, when employed irresponsibly, catastrophes might ensue. Knowing how to identify between various lighting signals and when to utilize each is vital, but you must also check that your automobile lights are operating as intended. This chapter will examine automotive lighting systems, their benefits, the various automobile lighting systems, as well as maintenance suggestions, and how to utilize your mobile device as an automotive lighting system inspection tool.

What is an Automotive Lighting System

An automotive lighting system offers the driver much-needed visibility on the road and vehicle operators, especially during night journeys, using lighting and signaling components positioned on the front, rear, sides, and even the top of the car.

The Importance of Car Lighting Systems

Each automobile contains internal and external illumination signals, from tail lights to headlights. These are vital for preserving road safety, notably at night. In addition, the road is lighted by external lighting, which increases automobile visibility and the driver's range of view.

The sight of traffic signs, markings, and barriers offered by the headlights increases our ability to respond swiftly to unforeseen impediments like road debris, animals, or potholes. Additionally, automobile lights warn other drivers of your presence on the road.

What are the Different Car Lighting Systems?

Getting acquainted with automobile lighting makes your working order more uncomplicated for you. Here are a few automobile lights that you should get acquainted with.

Headlight.

It offers great road sight, particularly at night or in low light.

Daylight-saving lighting

These lights are meant to boost a vehicle's visibility to other drivers; some drivers find them distracting in oncoming traffic.

Taillights

Its wiring enables it to switch on anytime the headlights are on and is only required to create a red light. Taillights notify other drivers of your presence and help them calculate your distance from their automobiles.

Dim light

Fog lights are set low so they won't refract and glare back at the car from the fog.

Brake lights

You may let other drivers know that you are slowing down or stopping by engaging the brakes.

Signal lights

These lights are intended to inform other cars that you will soon be taking a turn in the direction marked by the signal light and pedestrians.

Hazard lights

Hazard lights, often called flashers, should only be used to alert traffic concerns or other emergency circumstances.

Car interior lights
It lights up the vehicle's interior so the driver and passengers can swiftly examine maps or instructions and locate their items at night.

Advantages of Regular Lightning Vehicle Check-up

Here are the benefits you may obtain for a regular lightning check-up of your automobile.

1. **Safety:** The car's lights are essential for a vehicle to be noticed on the road. These lights perform various duties and are crucial for ensuring your safety, passengers, and the safety of other drivers.

2. **Visibility:** The primary purpose of a car's lights is to boost visibility inside and outside the vehicle. Regular lighting inspections will guarantee you can see the road effectively and supply the cabin with the right illumination.

3. **Communication:** Drivers use vehicle lights to communicate while driving. Regular lighting maintenance will guarantee that car lights are operating properly and maintain continual contact with other drivers.

4. **Information:** Dashboard lights and other inside automobile illumination transmit information crucial for every driver. These lights notify when anything is amiss with the vehicle's functioning and must be handled promptly.

Ways to Maintaining Your Car Lighting Systems in Order

Most of its important repairs involve little time and need extremely minor upkeep. The following ideas and recommendations assist in maintaining automobile exterior lighting.

Examine the Car

Regular examination is the first step in preserving the condition of your headlights, taillights, signal lights, and brake lights. A quick examination may be done to check for any indicators of flaws, malfunction, and uneven illumination.

Replace Any Damaged or Defective Components

Replacing or repairing broken, faulty, or damaged cars is ideal for everyone's safety and to conform to traffic rules and regulations.

Clean and Polish

It's vital to clean your automobile periodically, particularly the headlights. Regular cleaning and polishing assist in transforming murky headlights into clearer ones, offering you great sight even at night. Maintaining your car's outside lighting intelligently and properly enhances visibility and makes your driving safer.

CHAPTER 6: ELECTRICAL ACCESSORIES

Power Windows

The windows on automobiles which may be opened or shut with the use of buttons, are called power windows. Power windows were initially developed by Ford Motors in 1941. The first automobiles to have power windows were the Lincoln Custom and the Packard Custom Super 180.

Power windows have replaced the usual manual handles. It may be a built-in function or incorporated in automobiles, utilizing aftermarket equipment.

Features Of Power Windows

Keeping aside the common function of the windows being rolled up and down with

81

only the press of a button, the power windows come with certain distinctive characteristics that make them more user-friendly and handy. These traits may be summed up as follows:

Automatic Down: This is a fairly typical feature in virtually all power windows. This feature enables the user to make the windows go all the way down by merely touching the button once. The technique employs a circuit to measure the length of time that the switch was kept down. If the switch is kept down for less than a second, the window goes down all the way to hit the limit switch and stops there. In case the button is held down for a longer duration, the circuit assesses the time and stops rolling the window down as soon as the button is removed.

Automatic Up: This function is not very frequent since it has certain issues. The automatic up operates in the same manner as the automatic down. But, the automatic up function creates a danger. While the window goes up, if anything comes in the path of the window, such as a child's hand or a pet dog or cat's paw, it is extremely likely to damage them. As the window won't stop until it strikes the limit switch, it engages the danger of an accident. The only way out of this dilemma is the insertion of another circuit into the system, that can monitor the pace at which the window rolls up. As the speed is slowed down owing to an impediment, the circuit reverses the power back to the motor and the window goes down.

Courtesy Power-On: As mentioned previously, power windows operate only when the ignition of the car is on. But some automobiles have a courtesy power backup which is provided to the window circuit even after the engine is switched off. In case you forget to roll up your windows, this function saves you from the inconvenience of turning your ignition on again, only to pull the windows up.

Mechanism Of Power Windows

Power windows are controlled by switches and cables and are powered by battery or energy. Power windows do not work if the ignition of the automobile is not switched on. Unlike typical windows, power windows do not have manual handles. They do not work manually. The basic power windows normally have control of all four windows and may be operated by the driver.

This Is How Power Windows Work:

The door on the driver's side is controlled by a streamlined 20-amp electrical switch.
The electricity from the circuit breaker is sent to a contact point in the middle, where

the wirings of all four windows connect, via the window switch control panel.

Then the two ends of the power contact are linked to the battery or electric motor and the vehicle ground.
When a switch is pushed, one of the two ends gets separated from the vehicle ground and is connected to the power point in the middle. This distributes the electricity to the remainder of the windows.

However, power window systems utilized in current automobiles are quite sophisticated in terms of technology. These days most modern and high-end automobiles are fitted with several powered components including power windows, power doors, and power ORVMs (Outside Rear View Mirrors). In such circumstances, automobile manufacturers find it incredibly difficult to assemble all the wires in one. That is why all the wires are integrated into one module, and so all the controls are monitored. Since all the wires are combined into a single module, the power is delivered immediately to the central wire module and then to all four windows, with only the touch of a button by the driver.

Advantages Of Power Windows

Power windows are featured in practically all automobiles these days. In high-end automobiles, it is supplied as a standard and even in mid-range cars, it is at least given as an option. The electric windows remove the obstacle of the old handles which required a lot of effort and time to open or close the windows. The benefits of power windows may be summarized as follows:

1. It enables the driver to operate the windows with only the touch of his fingertips.

2. It permits those with hand injuries or other physical issues to simply manipulate the windows.

3. Drivers may effortlessly operate the windows even while driving.

4. The master power panel in the front also enables the driver to activate all the windows simultaneously, without leaving his position. This function is quite handy in case there are youngsters in the rear seat.

Disadvantages Of Power Windows

Although power windows provide a lot of benefits, they could also have some negatives at specific times. As it is powered by a battery or electricity, it is always coupled with the possibility of a failure or a breakdown. In case of a power window failure, one cannot open or close the window which may bring a lot of issues. Power windows could quit operating all of a sudden. Listed below are the most typical factors that could cause a malfunction or breakdown in power windows:

1. Many times, the window regulator also known as the window track, could cease operating. This causes a power window failure.

2. Power windows could also cease operating due to a faulty motor, a broken cable pulley, or a broken switch.

3. Worn-out window regulators could also be a factor behind a faulty power window.

4. With all the benefits that the power window system has to offer on the one hand and the minimal downsides on the other hand, it is very clear why most automotive manufacturers

provide power windows as a standard these days.

Power Locks

Late-model automobiles may be difficult, particularly concerning electric door locks. Whether your automobile has a keypad, keyless entry system, or standard power locks, locks may fail. And when they do, they may either leave you locked out of your car, or unable to lock your vehicle completely. The rising number of automobiles on the road today in Lee County, Florida that have central computers and power windows and locks has pushed drivers to seek professional technicians for repairs, rather than do it themselves. So, what precisely occurs when your car power lock unlocks?

The typical car gets locked and unlocked tens of thousands of times throughout the life of the vehicle. The device, the actuator, that is responsible for this continual locking and unlocking is meant to be dependable. Many current automobiles have four or five distinct methods to lock the doors and the car must keep track of the many signals that suggest that the car should be secured or opened.

With a physical key
Pressing the unlock button inside the car
Using a combination keypad on the outside of the door.

Pulling up the lock knob inside the automobile.

Using a keyless entry remote Control center signal.

LOCKING AND UNLOCKING
In certain model automobiles, the lock/unlock switch transmits power to the actuators and the door unlocks. In more complicated power lock systems, the body

controller determines the locking and unlocking. The body controller is your car's computer and it is in charge of various electrical processes inside your vehicle. The body controller reads lock/unlock orders from all signal sources whether they originate from a radio frequency, digital code, or radio transmitter from your key fob.

What Happens Inside The Car Door?

A power door lock actuator is normally positioned someplace below the latch. A connecting rod joins to the actuator and another one connects the latch to the manual knob that sticks out the top of the door. The mechanism enables for the rods to function in combination to power the actuator. Within the actuator, a small electrical motor drives a series of gears that drive a rack and pinion gear set that is connected to the actuator rod. The rotating action of the motor is translated into linear motion, and the lock is opened. A centrifugal clutch stops the action from occurring in reverse. The basic fact is that your vehicle's power door locks have numerous moving components and a lot may go wrong.

Everyone wants the newest automobile with all of the amenities, yet, the more features the vehicle has, the more that will need maintenance.

Air Conditioning And Heating Systems

Vehicle heating and cooling systems are crucial equipment that maintain the interior at the proper temperature and with acceptable air quality.

The heating system is equally vital for keeping nice temperatures during the winter months, even though the majority of people may only consider the air conditioning (AC) system when discussing automotive climate control. It's vital to appreciate the basic pieces, how the systems operate together, and how they're maintained to comprehend how these systems perform.

Heating And Cooling Systems Components

Several sections make up the heating and cooling systems of automobiles, which work together to maintain the right cabin temperature. The heating and cooling coils, compressor, evaporator, condenser, fans, and hoses are the most critical pieces.

The coils used for heating and cooling are in charge of extracting heat energy from or releasing heat energy into the environment. The heater core or evaporator core receives the heat energy via a network of pipes and radiators that are attached to the engine. The air is cooled or heated before it enters the cabin via the heater core or evaporator core, which is coupled to the air conditioning system.

The air conditioning system's heart is the compressor. Compressing refrigerant gas boosts its temperature and makes it more able to absorb heat from the surrounding air. The condenser cools and turns the hot refrigerant gas into a liquid after being pushed there by the compressor. The liquid refrigerant is then delivered to the evaporator, where it is turned back into a gas by absorbing heat from the cabin air. The method that cools the cabin air is this one.

The air circulation in the heating and cooling system is done by the fans. The evaporator and condenser are coupled to the fans, which draw air through these sections to help the air conditioning system chill the cabin air.

The hoses are in charge of delivering refrigerant gas among the different sections of the air conditioning system. Between the engine and the cooling coils, the hoses also transfer engine coolant.

The Operation of Heating and Cooling Systems

The heating and cooling systems combine to manage the interior temperature of the automobile. The evaporator cools the cabin air by collecting heat from the air inside the cabin, while the heater core heats the cabin air by transferring heat from the engine coolant.

The compressor compresses the refrigerant gas when the air conditioner is powered on, elevating its temperature. The condenser is where the hot refrigerant gas is delivered after cooling and becoming a liquid. In the evaporator, where it takes up heat from the cabin air and converts it back into a liquid, the refrigerant is then transported. The cabin air is cooled by this technique. The fans pull the cooled air into the cabin and aid in moving the air through the system.

As soon as the heater is triggered, the engine coolant is injected into the heater core, where it warms the interior air. The fans serve to spread the heated air throughout the cabin.

UPKEEP OF HEATING AND AIR CONDITIONING SYSTEMS

To make sure that heating and cooling systems are running correctly, they must be constantly examined and maintained. Checking the fluid levels in the system and ensuring sure the hoses are in great condition is crucial. Additionally, search for leaks or indications of wear and tear on the compressor. Additionally, it's vital to frequently clean the condenser and evaporator to minimize the collection of dirt and debris, which may impair the system's efficiency.

Audio Systems

What is a vehicle audio system?

Three key components make up your automotive audio system, connected by power and audio wires:

Head Unit, Amplifier and Speakers

Depending on your arrangement you may additionally require a crossover.

THE HEAD UNIT

The head unit is known by several other names including stereo receiver, vehicle stereo, and car radio. The head unit is situated prominently in the dashboard and offers an interface for the automobile audio system, including buttons, system controls, and in some instances displays. The head units have their amplifiers for powering the speakers but may include alternative outputs for more powerful, stand-alone amplifiers.

It is the core portion of the audio system: adjusts the sound level controls the treble and bass controls.
regulates access to the various audio sources In addition to functioning as the brains of the audio system, certain head units are also incorporated into infotainment systems. These head units often have big LCDs, and they are frequently capable of showing navigation data, regulating temperature settings, and performing other activities.

THE AMPLIFIER

The amplifier is the second important component of all automotive audio systems. The amplifier must raise the strength of the audio signal generated by the head unit, so it is powerful enough for the speakers to output.

Without a power amplifier, the audio signal will be too weak to physically move the speakers and make sound.

If you want a louder more powerful sound that the head unit's in-built amplifier can deliver you will need an extra amplifier. More power also enables you to attach more speakers if you want completely immersive vehicle audio.

The 'maximum power 'is the greatest the amplifier can go, while the 'rated power' is the usual operational output. The higher the ratings are, the more power it can send out to the speakers and the louder it can go.

Amplifiers have a varying number of channels from mono (single channel) up to 5 channels. You will require one channel for each speaker on your audio system.

When you place an amp in your automobile, it will require sufficient space around it to enable the heat to escape - the more powerful they are, the more heat they may create. For security purposes, it's also ideal to mount it in a concealed area. The most typical mounting sites are below the front seats, or within the boot.

THE SPEAKERS

Speakers are the last component of the basic automotive audio system.

A speaker receives an audio signal from an amplifier, and the electrical energy of the signal is turned into mechanical energy that causes a cone to move back and forth. That movement or vibration displaces air which produces the sound waves we perceive.

Coaxial and component are the two kinds of speakers that may be utilized to create or update your car's audio system.

Component speakers are superior in terms of sound quality, while full-range speakers are less costly and simpler to install. Component speakers give a larger option for customization. In addition to superior sound quality, component speakers may be independently positioned to produce the optimal soundscape for a specific car. If sound quality is more essential than cash or time, component speakers are the way to go.

Coaxial Speakers

The most popular automotive audio systems feature a head unit (with a modest amp within) and four coaxial speakers. Frequently known as full-range speakers because they reproduce a greater range of audio frequencies from a single device.

A coaxial speaker comprises a woofer that generates low- and mid-range sounds and a tweeter that sits above the woofer providing high-range sound. Superior coaxial sets will include additional tweeters to better recreate the high-range sound.

Coaxials feature a basic crossover built-in that takes the input signal and separates it into different bands of high, medium, and low frequencies. This assists in assigning the signals to various portions of the speaker. They come in a few different band arrangements.

The quality of coaxial speakers ranges from inexpensive ones built with foam and paper to made of fabric and foam construction, to rubber, metal, silk, and synthetic materials in the finest automotive audio.

Even the top brands of coaxial speakers have limits because of the inherent difficulty of coupled components. Having the tweeters fastened to the speaker reduces the separation of frequencies and allows for genuinely excellent quality music.

Component Speakers

Component speakers are specially constructed for either bass, midrange, or high frequencies, and therefore, will provide you with the greatest sound quality available.

You may add as many speakers as you want in as many configurations as you desire. Not only will you have a significantly greater volume than coaxial speakers are capable of, but the quality of sound will be noticeably improved. This is because even at high levels, the independent crossovers will make sure the speaker components are not being pushed beyond their capacities.

With a greater separation of the audio spectrum, individual instruments stand out considerably more. When you have a decent component speaker system, it sounds like the musician is playing right in front of you. The range of human hearing is around 20 to 20,000 Hz, and that spectrum is often split into a few categories when it comes to speaker technology. Component speakers each handle a specific section, or component, of that spectrum. The highest frequencies are provided by tweeters, the lowest by woofers, and mid-range speakers sit between those extremes.

Tweeters

Tweeters span the upper end of the audio spectrum from roughly 2,000 to 20,000 Hz and frequently play a vital role in filling out an audio environment. These speakers are called after the high-pitched tweeting of birds. Tweeters generate extremely directed sound and, hence are best positioned at ear level.

Mid-range

The middle range of the auditory spectrum consists of sounds that fall between 300 and

5,000 Hz, hence there is some overlap between mid-range speakers and tweeters.

Woofers

Woofers handle deep bass which falls in the region of around 40 to 1,000 Hz. Woofers create low-frequency sound, therefore it's OK to put them down in the door panels.

There are also a few specialist component speakers that give enhanced fidelity at the extremes of the audio spectrum.

Super Tweeters

These speakers are occasionally capable of generating ultrasonic frequencies that are above the typical range of human hearing, and their lower ends are substantially higher than the 2,000 Hz that standard tweeters handle. That enables super tweeters to generate higher-frequency sounds without any distortion.

Subwoofers

Like super tweeters, subwoofers are meant to offer greater quality sound at one end of the audio spectrum. Consumer-grade subwoofers normally function in a range from 20 to 200 Hz, however, professional sound equipment might be restricted to frequencies that are below 80 Hz. Subs pump out all the low-end sounds such as bass, drumming, and deep voices.

A typical component system contains a woofer, tweeter, and external crossovers - all of which are particularly built to sing in tune.

Crossovers

If you want a component audio system then youpn9 will need more than one crossover. A crossover sends audio frequencies to the right target speaker — ensuring sure only high-frequency sound reaches the tweeters and low-frequency sound finds its way to the subwoofer.

There are two basic kinds of crossovers, passive crossovers, and active crossovers — each of which is better suited to distinct situations:

Passive Crossovers: Passive crossovers exist between the amplifier and the speakers to filter out undesired frequencies.

However, there is a certain element of inefficiency that is inherent in passive crossovers.

Active Crossovers: Active or electronic crossovers reside between the receiver and the amplifier, where they send frequencies to the right speakers before the amplification process.

An active crossover needs a power supply, but they don't waste electricity by filtering out amplified sounds the way passive crossovers do.

Whether you use active or passive crossovers, you'll need something to restrict undesirable frequencies from reaching the speakers.

CHAPTER 7: TROUBLESHOOTING AUTOMOTIVE ELECTRICAL SYSTEMS

Troubleshooting electrical difficulties may be a laborious process, but it does not have to be if you keep a few fundamental points in mind: Every circuit needs a power source; most electrical devices need a minimum voltage to perform efficiently; and all circuits require continuity. Consequently, most electrical faults are caused by low voltage (or no voltage), excessive resistance, or a loss of continuity.

SAFETY CONSIDERATIONS
Safety is always a big consideration when working on automotive electrical systems. Except for the high voltage side of the

ignition system, and the high voltage battery and circuits in hybrid autos, there is little possibility of being shocked. Twelve volts (12v DC) isn't sufficient to be felt.

The concern includes inadvertently shorting out a hot circuit and destroying the wiring, PCM, or other onboard electronics, or triggering a fire.

CAUTION: If your automobile is a hybrid with a high-voltage battery, there is a possibility of being shocked if you come into direct contact with the high-voltage battery, cabling, or other hybrid components.

CAUTION: When doing electrical repairs or replacing electrical or electronic components, the battery should ALWAYS be disconnected to eliminate any potential of generating an unintended short. Disconnecting the battery will cause most PCMs to lose their learned settings. This might cause driveability issues or require an extraordinary "relearn" methodology with a sweep instrument, so to stay away from this sort of problem utilizing a 9-volt "memory saver" that plugs into the vehicle's power receptacle (cigarette lighter) to maintain voltage to the battery, or connect a 9-volt alkaline battery to the PCM power supply.

Basic Automotive Electrical Circuit Checks

All electrical circuits require voltage to operate the components attached to that circuit. So assuming there is no voltage, there is no capability.

The first order of business when troubleshooting electrical faults, therefore, is to check for the presence of voltage at the load point in the circuit.

The load point is the device that the circuit is meant to power, such as a light bulb, wiper motor, blower motor, idle stop solenoid, or whatever. And, all you need to fast check is a voltmeter or a 12-volt test light that lights when there is energy. A voltmeter is a great instrument for this purpose as it will give you an accurate measurement, but a test light is OK for making rapid voltage checks.

Electrical test light

Using a test light is a simple approach to check for voltage, although a voltmeter is more accurate.

Suppose you detect no voltage at the load site. Ah ha, you have unearthed your first suggestion about the matter. Check the fuse, fuse link, or circuit breaker that secures the circuit, or the power relay that provides energy to the circuit.

If the problem is a blown fuse, replacing the fuse may restore power briefly, but unless the underlying cause for the overload is found and corrected, your "repair" likely will not survive. Whatever you do, do not replace a fuse of bigger capacity.

A bigger fuse might have the option to endure a more prominent burden yet the wiring and the remainder of the circuit can't. A circuit configured for a 20 amp fuse is meant to handle a maximum of 20 amps. Period.

A deficient electrical switch or an open hand-off will have a similar impact as a blown circuit.

Circuit breakers are widely deployed to protect circuits that may suffer brief periods of overloading such as an A/C compressor clutch.

The fastest technique to investigate a circuit breaker is to bypass it using a jumper wire. Your jumper wire should have a replaceable inline fuse to preserve the circuit from injury. Utilize a wire of no more prominent limit than what the actual circuit utilizes. If you don't have any idea, utilize a 5- or 10-amp circuit to be protected. On the off chance that the circuit works when you sidestep the electrical switch, you have segregated the issue.

Replace the circuit breaker

This same easy test may also be conducted to check a suspect relay. A relay is nothing more than a remote switch that utilizes an electromagnet to close a collection of contact points. When the relay magnet is fed with power, the points shut, and the battery voltage is routed via the main circuit. Relays are frequently used in circuits to minimize the quantity of wire that is required and to lessen the current that flows through the primary control switch. Thus, a relatively low amperage (make that affordable) switch, timer, or sensor may be employed to turn a substantially bigger capacity relay on and off.

Voltage Checks For Car Electrical Problems

Every electrical item also requires a precise degree of voltage to function. A light bulb will illuminate with diminishing brightness as the voltage drops.

However, for certain parts, there is a limited voltage underneath which won't work by any stretch of the imagination.

A starting motor may crank the engine more slowly with decreasing voltage but, if the battery voltage is so low, it may not crank at all. The minimum threshold voltage is particularly crucial for such components as solenoids (which require a specific amount of voltage to overcome spring resistance), relays, timers, buzzers, horns, fuel injectors (which are solenoids, too), and most electronics (the ignition module, computer, and radio).

Checking the load point for full battery voltage can tell you whether or not the appropriate voltage is getting through, and to achieve so you need a voltmeter. The battery itself should be at least 70 percent charged and display 12.43 volts or higher (12.66 volts is entirely charged). The output of the charging system should also be checked, and be roughly 1.5 to 2.0 volts higher than the battery base voltage (around 14 to 14-1/2 volts).

Assuming the battery is all right, your voltmeter ought to peruse inside 1 volt of battery voltage at the circuit load point in some random circuit.

Low circuit voltage is typically caused by excessive resistance at some point in the wire. Usually, this suggests a loose or corroded connection, a malfunctioning switch or relay, or weak ground. To find the location of excessive resistance, use your voltmeter to run a "voltage drop test" at many spots throughout the circuit. If the voltmeter exhibits a reduction of more than 0.4 volts across any connection, switch, or ground contact, it implies danger. Ideally, the voltage loss should be no more than 0.1 volts.

If low voltage is found in numerous circuits, do a voltage drop test between the battery terminals and engine/body ground straps. Loose or corroded battery cables and ground straps are a significant cause of voltage-related troubles. Clean and fix the battery cables as well as ground straps, if important.

Sometimes insufficient wiring could create low voltage. It is not something you will see in many original equipment wiring circuits, but it is a regular blunder that is done in many do-it-yourself wire installations for aftermarket accessories. The larger the amp load in the circuit, the bigger the requisite gauge size for the wire.

Electrical Continuity Tests

Each electrical circuit requires a total circuit to work. Voltage to the load will not do any good unless there is also a complete ground connection to the battery. The ground route in the case of all metal-bodied cars is the body itself. In plastic-bodied autos, a separate ground wire is necessary to link the load to the chassis. In every case, a faulty ground connection has the same effect as an open switch. The circuit is not complete consequently current does not flow.

To check wire continuity, you need an ohmmeter or a self-powered test light. An ohmmeter is a preferable alternative as it provides the exact amount of resistance between any two test locations. A test light, then again, will sparkle when there is coherence however the force of the bulb might shift depending on how much opposition is in the circuit. But it is OK to do rapid checks.

Never use an ohmmeter to determine resistance in a live circuit. Make sure there is no voltage in the circuit by unplugging it from its power source, pulling the fuse, or checking downstream from the circuit switch or relay. Ohmmeters cannot resist average battery voltage and, should you unintentionally complete a circuit via the meter, you may kill your meter.

Ohmmeters are fantastic for measuring circuit resistance but you have to be cautious when testing electrical components. An ohmmeter functions by sending a moderate voltage via its test leads, and this voltage may be enough to kill some electrical components (such as the oxygen sensor). Special high impedance 10,000 mega-ohmmeters should be employed for electronics testing.

Finding Electrical Faults In Your Car's Wiring

Now that we have covered several essential troubleshooting techniques, what is the best strategy to find an electrical fault fast? It depends on the nature of the circumstance.

For a "dead" circuit, the first thing to look for is a voltage at the load point. Utilize your voltmeter or 12-volt test light to check for voltage. If there is voltage, the problem is either a weak ground connection or the component itself has failed. Check the ground connection with your ohmmeter. If the ground connection is OK, the fault resides inside the component. If there is no voltage in the "hot" line to the component, then the issue is in the wiring. Traceback via the fuse panel (or relay or circuit breaker) until you detect power. Now hunt for an open or short that is stopping the current from reaching its appropriate destination.

Next comes bad connectivity. The resistance created by a loose or corroded connection might induce a voltage drop that can have an adverse influence on circuit components. An ohmmeter may be used to evaluate non-powered circuit connections for extra resistance, but a better technique is to use a voltmeter to check for a voltage drop across a connection.

Voltage Drop Test Using A Voltmeter

The voltmeter leads are connected on each side of the circuit component or connection that is being checked. If a connection is loose or rusty, it will generate resistance and give a reading on the voltmeter. As noted earlier, a voltage loss of more than 0.4 volts suggests danger, and ideally, it should be 0.1 volts or less.

The worst form of electrical problem to tackle is an intermittent one. Everything works properly at the shop but as soon as the customer takes the car back it starts to act up again. An intermittent open or short is typically the effect of something heating up and breaking (or generating) contact or something that is loose and is making periodic contact.

Loose or corroded connections and switches are usually responsible for this kind of problem, so try jiggling the wires and circuit switch to see if it impacts circuit voltage or resistance. A wire that is rubbing and has chaffed away part of its insulation can make intermittent contact generating a short, therefore wriggling suspicious wires can generally uncover the problem.

Temperature-sensitive intermittent shorts or openings may be tough to identify as you normally have to reproduce the specific circumstances that cause them to emerge. Sometimes you may guess what is happening by the nature of the circumstance. But it is typically more enjoyable (and soothing) to replicate the situation so you know for sure what is wrong.

When does the problem occur? Does it merely happen when the engine is hot or after the circuit has been on for a length of time? Using a hot air pistol or hair dryer to heat wires, connectors, switches, and relays may sometimes aid in detecting damaged components.

Environmental conditions often wreak havoc with electrical systems, too. Road splash or water escaping from a crack in the cowl, behind the windshield, or around a grommet may sometimes short out a circuit. Look for evident indicators of corrosion or leakage, and if you find none examine the condition of nearby weather seals.

A final note on rectifying electrical faults: When splicing wires do not just twist them together and place electrical tape over the connection. Use a solderless crimp-on connection, or twist the wires together, solder them, and then use shrink-wrap electrical insulation tubing to seal the repair.

CHAPTER 8: ADVANCED AUTOMOTIVE ELECTRICAL SYSTEMS

Electronic Fuel Injection Systems

What is an Electronic Fuel Injection System (EFI)

A system aimed at improving the fuel/air ratio entering a vehicle's engine is termed an Electronic Fuel Injection System. EFI System has nearly fully replaced the use of carburetors.

Carburetors are fantastic for performance, but owing to their vague nature, they can't create big horsepower, get a respectable gas economy, and pass an emission test, all with the same tune, they also have numerous mechanical components that may get sticky over time. This means they were more maintenance-intensive, with a carburetor rebuild typically being part of a normal maintenance cycle.

OEMs turned to EFI to handle their complicated emission concerns. The first EFI was mainly simply processor-controlled carburetors coupled to an oxygen sensor and throttle position sensor, all wired to an Electronic Control Unit.

The electrical Fuel Injection System comprises electrical components and sensors. It needs to be maintained clean and properly adjusted to improve the engine's strength and efficiency and to cut down gas usage.

Types of Fuel Injection

To comprehend the notion better, we first have to understand the forms of Fuel Injection. The Fuel Injection types utilized in contemporary automobiles are:

- Single-Point or Throttle Body Injection

- Port or Multi-Point Fuel Injection

- Sequential Fuel Injection

- Direct Injection

- Single-Point or Throttle Body Injection

The original and basic sort of Fuel Injection was the Single-Point injection
Here the carburetor is supplanted by a couple of fuel injector spouts in the choke body, which is the throat of the motor's air consumption complex.

Single-Point Injection was a stepping stone to the more sophisticated multi-point system for certain manufacturers. They are inexpensive and easy to service.

Port or Multi-Point Fuel Injection

In Multi-point Fuel injection a distinct injector nozzle is allocated to each cylinder, immediately outside its intake port, owing to which the system is also called a Port Injection System. When the fuel vapor is blasted near the intake port it makes sure that fuel is sucked entirely into the cylinder.

The key advantage is that MPFI meters fuel more accurately than the TBI designs do. It is superior at obtaining the optimum fuel/air ratio and enhancing all connected elements. Also, it practically prevents the chance of the gasoline from condensing or to be accumulating in the intake manifold. The TBI and carburetors are created in such a manner that the intake manifold transmits the engine's heat which is the measure to evaporate liquid gasoline.

It is not essential for engines that are equipped with MPFI where the intake Manifold may be manufactured out of lighter-weight material, even plastic. MPFI technology results in enhanced fuel efficiency. Standard metal intake Manifolds have to be positioned on top of the engine to transmit heat, however, in the case of MPFI it may be positioned more imaginatively, providing the engineers with design freedom.

Sequential Fuel Injection

Sequential Fuel Injection, commonly known as the Sequential Port Fuel Injection (SPFI) or Timed Injection, is a kind of Multi-Port Injection. Albeit the MPFI has various injectors, they all splash their fuel simultaneously or in gatherings. This may result in "hanging around" of the gasoline in a port for as long as 150 milliseconds at the time of engine idle.

It may not seem like much, but it's sufficient of a constraint that the engineers fixed it i.e. Sequential fuel injection stimulates every injector nozzle independently. They are essentially timed like spark plugs and spray the gasoline exactly before or as their intake valve opens. Although sounds like a little step, the efficiency and emission gains occur in incredibly small dosages.

Direct Injection

Direct Injection injects gasoline directly into the combustion chambers, past the valves. Direct Injection system is widespread in diesel engines and is starting to come up in gasoline engine designs, commonly named DIG for Direct Injection Gasoline. Fuel proportioning is still more exact than in the other injection method.

Direct Injection provides engineers an extra option to regulate exactly how combustion happens in the cylinders. The study of engine design scrutinizes how the fuel/air combination swirls about in the cylinders and how the explosion travels from the ignition point. Direct Injection may be employed in low-emission lean-burn engines.

Components of Electronic Fuel Injection System

Components of the Electronic Fuel Injection System include:

- Sensors

- Electronic Control Unit (ECU)

- Fuel Injectors

- Fuel Pump

- Sensors

131

Sensors are positioned at several areas of the engine and their role is to give out information to the ECU. The following sensors are used:

- Engine Temperature Sensor

- Intake Temperature Sensor

- Exhaust Temperature Sensor

- Engine Speed Sensor

- Throttle Position Sensor

Probe which is responsible for the measurement of the fuel content in the fuel/air mixture.

Actuators are components that collect information from ECU and they function in the feeding system, adjusting the amount of gasoline that the engine receives.

It employs the following actuators:

- Fuel Injector

- Spark Plug

- Throttle

- Electronic Control Unit

The Electronic Control Unit is responsible for the measurement of the sensors and the calculation of the action for each actuator with consideration of time limits. The temporal restrictions of the system are imposed by the characteristics of the internal combustion engine to be managed.
It is specified that a rotation of the engine i.e.
$360°$ is performed once in each 5 miniature seconds for 12,000 rpm. The actuator of the choke valve deciphers the place of $0°$ as a beat of 1 milli second and $90°$ as a beat of 2 milliseconds, inside a period of 25

milliseconds. Considering this timing limitation, the reading of the sensors and the computation of the acting times for the actuators should be done in at most 15 microseconds.

The "Check Engine" Light (or "Service Engine Soon" Light) on the console comes on during the scan and turns off when all sensors are working.

Fuel Injector
It aids in pumping gasoline into the engine's intake ports.

Fuel Pump
It aids in pumping gasoline from the vehicle's fuel tank to the engine and delivers fuel to the Fuel Injection System at higher pressure.

How the EFI System Works

The Fuel Injection System comprises various sensors positioned all over your car. Every time you start your car, the Electronic Control Unit (ECU) analyzes every one of these sensors to verify their performance.

The "Check Engine" Light (or "Service Engine Soon" Light) on the console comes on during the scan and turns off when all sensors are working.

The sensors continually measure the values of multiple parameters including air pressure, air temperature, throttle angle, air density, fuel temperature, fuel pressure, oil pressure, coolant temperature, exhaust temperature, crankshaft angle, timing, engine rpm, speed, etc.

All these data are processed by the ECU (Electronic Control Unit) to determine the length of time the gasoline Injectors are open, pouring gasoline into your engine's intake ports. The injectors normally are open just for a few milliseconds at a time.

A fuel injector comprises a spout and a valve. The power to inject the gasoline comes from a gasoline Pump or Pressure Container positioned far back in the fuel supply. The fuel passing through the system is atomized by aggressively forcing it through a tiny nozzle under extremely high pressure.

Applications of Electronic Fuel Injection System

The applications include:

1. EFI System includes the state-of-the-art control software

routine for pollution, fuel economy, and performance needs.

2. The System also contains a technology of Smart Ignition for control of the ignition system, thus offering OEMs the freedom to achieve best-in-class fuel consumption

Advantages of Electronic Fuel Injection System

The benefits are:
1. Enhancement of volumetric efficiency of the engine

2. Direct fuel injection into the cylinder avoids manifold wetness

3. Good atomization of fuel even at low rpm since atomization is independent of cranking speed

4. Less banging due to the better atomization and vaporization
5. Ice development at the throttle plate is eliminated

6. Fuels with low volatility may be utilized as distribution is independent of vaporization.

7. As the fluctuation of the fuel/air ratio is very minimal it results in excellent engine performance

8. The height of the engine may be reduced since the location of the injection unit is not very crucial

Disadvantages of Electronic Fuel Injection System

The downsides include:
1. High maintenance cost

2. Difficulty in servicing

3. Possibility of breakdown of some sensors

Anti-lock braking systems (ABS)

With ABS, when you use the brakes, the speed sensors measure the decreasing rotation of the wheels. When the brakes are ready to halt spinning, they transmit a signal to the electronic control unit (ECU).

The ECU partly releases the brake pads from the wheels using valves and pumps, enabling the wheel to continue revolving. With ABS, the wheels may continue moving, helping you to keep control over the automobile in harsh braking conditions.

Without ABS, the wheels would halt or lock up immediately after the brakes are applied and the car would slide owing to the translational velocity of the wheels. In this instance, owing to skidding the car would traverse a reasonably long distance and you won't be able to manage the vehicle because you can't steer the locked wheels. Moreover, if the left and right wheels of the vehicle are on distinct traction surfaces, the use of brakes creates differing frictional forces on the wheels. This causes the car to spin uncontrollably.

Advantages of Anti-Lock Braking System or ABS

1. An anti-lock braking system minimizes the braking distance when the necessary pressure is delivered to stop the automobile.

2. Since wheels don't lock up when stopping the car, it reduces uneven tire wear.

3. ABS shares some fundamental infrastructure with the Traction Control System (TCS), which makes it simpler to install the traction control system on automobiles in factories.

4. Decrease the wearing of brake pads and brake discs.

5. In case of strong braking, you may guide the automobile around obstructions.

Disadvantages of Anti-Lock Braking System or ABS

1. Variable braking distances owing to various kinds of surfaces.

2. The elaborate combination of ECU and sensors increases the complexity

3. Expensive to maintain

Traction Control Systems (Tcs)

The traction control system (TCS) is an active car safety element. It avoids loss of traction of the wheels operated on roadways. It becomes activated when the engine torque and throttle input do not match the surface you are driving.

But, what does traction control mean? Well, it is a technology that restricts the power given to the wheel to put traction on the automobile wheels – making them stop spinning. It uses all the traction available on the road as the car accelerates on low-friction road conditions.

What Is a Traction Control System?

The TCS is an active vehicle safety device that is typical in current vehicles. Being a supplementary function of a vehicle's electronic stability control (ESC), the onboard system kicks in when the acceleration gets up. It stops the tires from sliding as the automobile accelerates up.

Traction control is beneficial in poor weather circumstances including rain, ice, or snow, and on slippery terrain that gives little to no traction. Drivers have to feather the throttle pedal on antique automobiles with no TCS mechanism to prevent the wheels from crazy spinning on rough roads.

It helps them to gain speed without losing grip. But, the latest automobiles equipped with this technology enable drivers to accelerate under control by restricting the power supply to stop the wheel slide.

The device helps the driver exercise greater control over their automobile. It decreases the danger of losing control of the rear end when accelerating and oversteering whilst driving around a corner. ESC is a computer-operated function in contemporary cars. It contributes to the stability upon finding out the loss of traction and then minimizing it.

How Does Traction Control Work?

Unless you are driving an ancient automobile, the odds are that it has a traction control system. It is in operation and aids with safe driving without you recognizing how it works.

Well, the technique is very clear if you pay attention. Its function is relatively similar to the anti-lock braking system (ABS) and the components of the two systems are likewise comparable. The TCS contains a wheel speed sensor that monitors the rotation speed of either the front or all four wheels. The hydraulic modulator powers the brakes while the ECU (electronic control unit) checks the data from wheel speed sensors and steps in to direct the hydraulic modulator if required.

In current automobiles, the ABS and TCS are regarded to be one unit since the latter is an add-on to the ABS. The ECU monitors the wheel action. It evaluates if somebody is spinning quicker than the others – which signals that the particular wheel is losing traction. In such a situation, the ECU steps in to tone down the rotation of the affected wheel. It directs the hydraulic modulator to pump the brake in fast succession to that particular wheel.

Some TCSs manage the wheel spinning by lowering the engine power to the wheels that are ready to lose traction. Once the situation returned to normal, the system continued to its usual function - monitoring the wheel and rotational speed.

When the system reduces the engine power to manage the sliding wheels, you may sense a throbbing feeling via the gas pedal. It is the influence of the TCS causing the engine to alter the wheel rotation speed. So, don't be alarmed if you sense a vibration while driving on a slick road.

The Right Time to Utilize Traction Control in Automobiles

The TCS is a typical safety element in current automobiles. It becomes active when the engine starts. The mechanism interferes when there is a difference between the wheel's rotation speed and the vehicle's speed.

The technique is advantageous to any sort of automobile, independent of the engine they have. But, it causes more damage than help in certain cases. For example, when you drive the vehicle from a stop position in snow or sand, all the wheels slide, causing the TCS to decrease the engine power too much. As a consequence, going forward on such surfaces becomes quite difficult.

Most models offer an option for shutting off the TCS in such challenging situations. Some autos don't allow to disable it altogether. But, they give a setting for snowy or slippery situations or an option to enhance the wheel spin if required. You have to switch off the system or apply the extra settings depending on the weather or the kind of surface you are traveling on.

Most models offer an option for shutting off the TCS in such challenging situations. Some autos don't allow to disable it altogether. But, they give a setting for snowy or slippery situations or an option to enhance the wheel spin if required. You have to switch off the system or apply the extra settings depending on the weather or the kind of surface you are traveling on.

CONCLUSION

Overall, this book provides a wealth of information and practical advice for those looking to understand and work with automotive wiring and electrical systems. The author's attention to detail and clear explanations make it accessible to readers of all skill levels. With this guide in hand, readers will have the confidence and knowledge to tackle any electrical issue they may encounter. From understanding electrical principles to diagnosing and repairing faults, this book covers all aspects of automotive wiring and electrical systems. As the final pages turn, readers will feel empowered and equipped to handle any electrical challenge that comes their way. This book truly is the complete guide to automotive wiring and electrical systems, and is an essential resource for any car enthusiast or mechanic.

www.ingramcontent.com/pod-product-compliance
Lightning Source LLC
Chambersburg PA
CBHW072210290526
45794CB00004B/1716